Oh Taste - 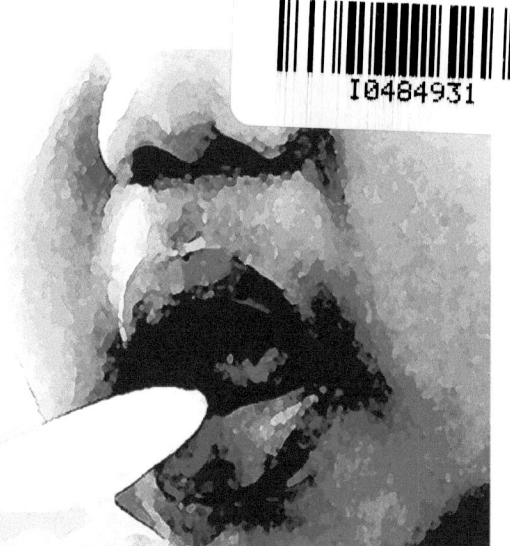 can You See

What Melanin Does For The Senses

by Stephanie Abena Kaashe

c. 2014

DEDICATION

This book is dedicated to the growing efforts of teaching melanin science to our children. Knowledge is power when you understand how to use it.

Oh taste can you see
the melanin in me
helping my senses
helping my senses

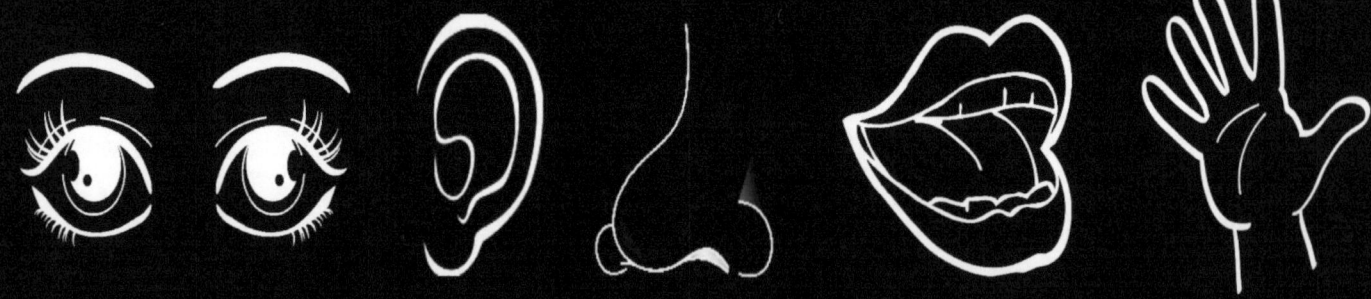

1

Oh taste can you see
the melanin in me
helping my senses
do amazing things

2

Melanin
helps my senses
to do their best work

3

There are five altogether let's start with the eyes first

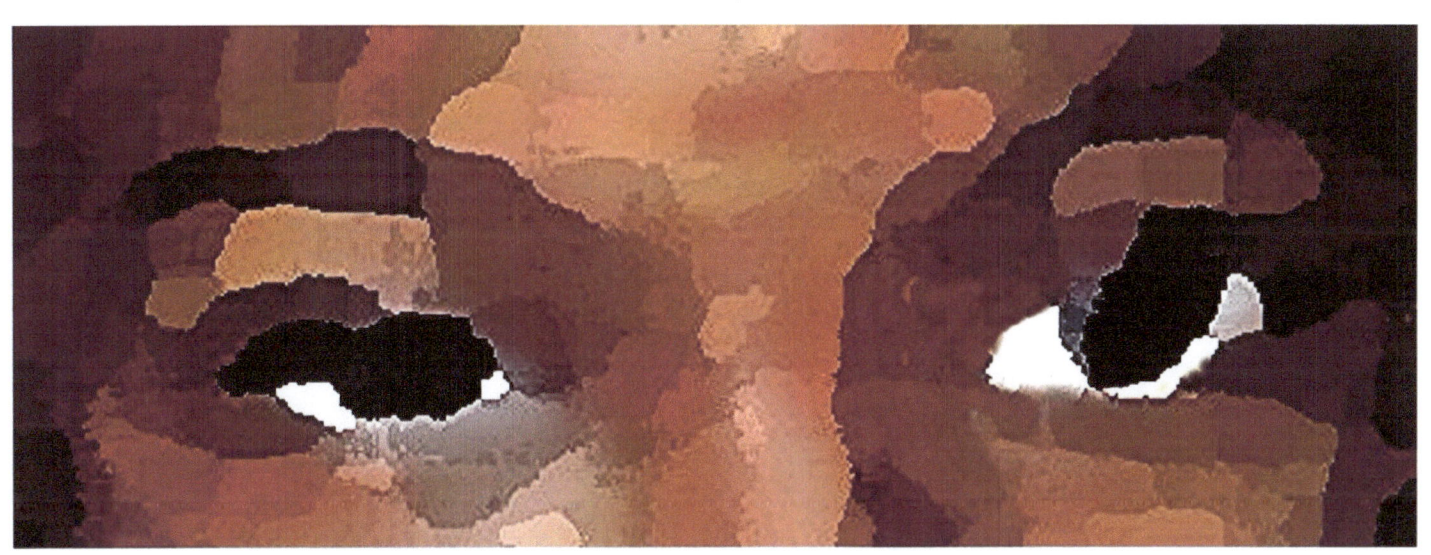

With melanin I can see all the colors in the rainbow

5

From sunshine yellow to deep indigo

6

With my nose I can smell every flower and scent

All the herbs in the garden sweet basil and mint

8

Melanin in my ears
helps me hear every sound

The birds in the air

The crickets on the ground

With my tongue I can taste
bitter, spicy or sweet

**Dad's winning chilli
and mom's special
treats**

Melanin flows
through my fingers and toes

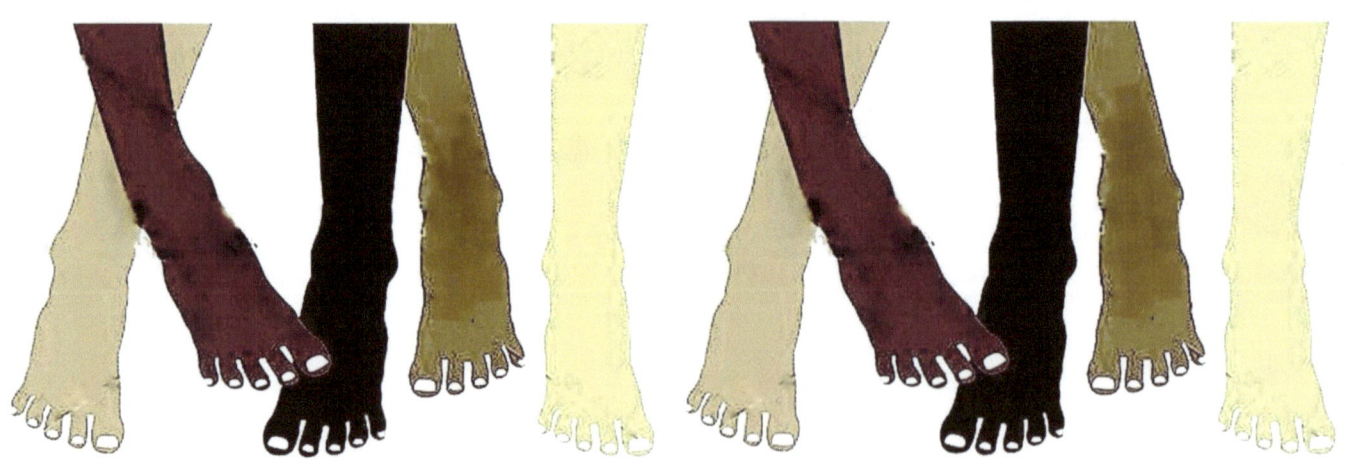

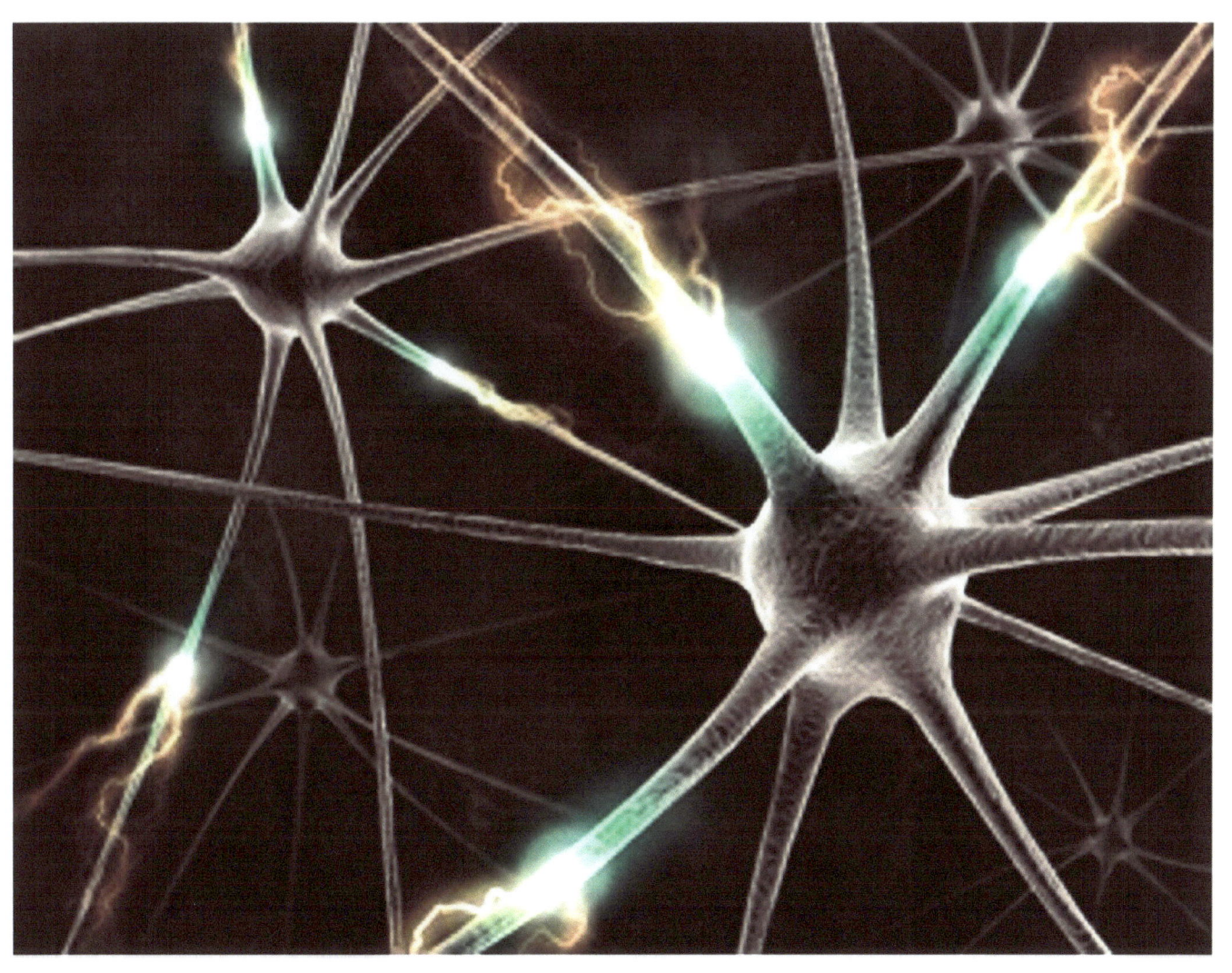

**Traveling through nerve endings
I can feel where it goes**

14

**Melanin works wonders
so my senses do much**

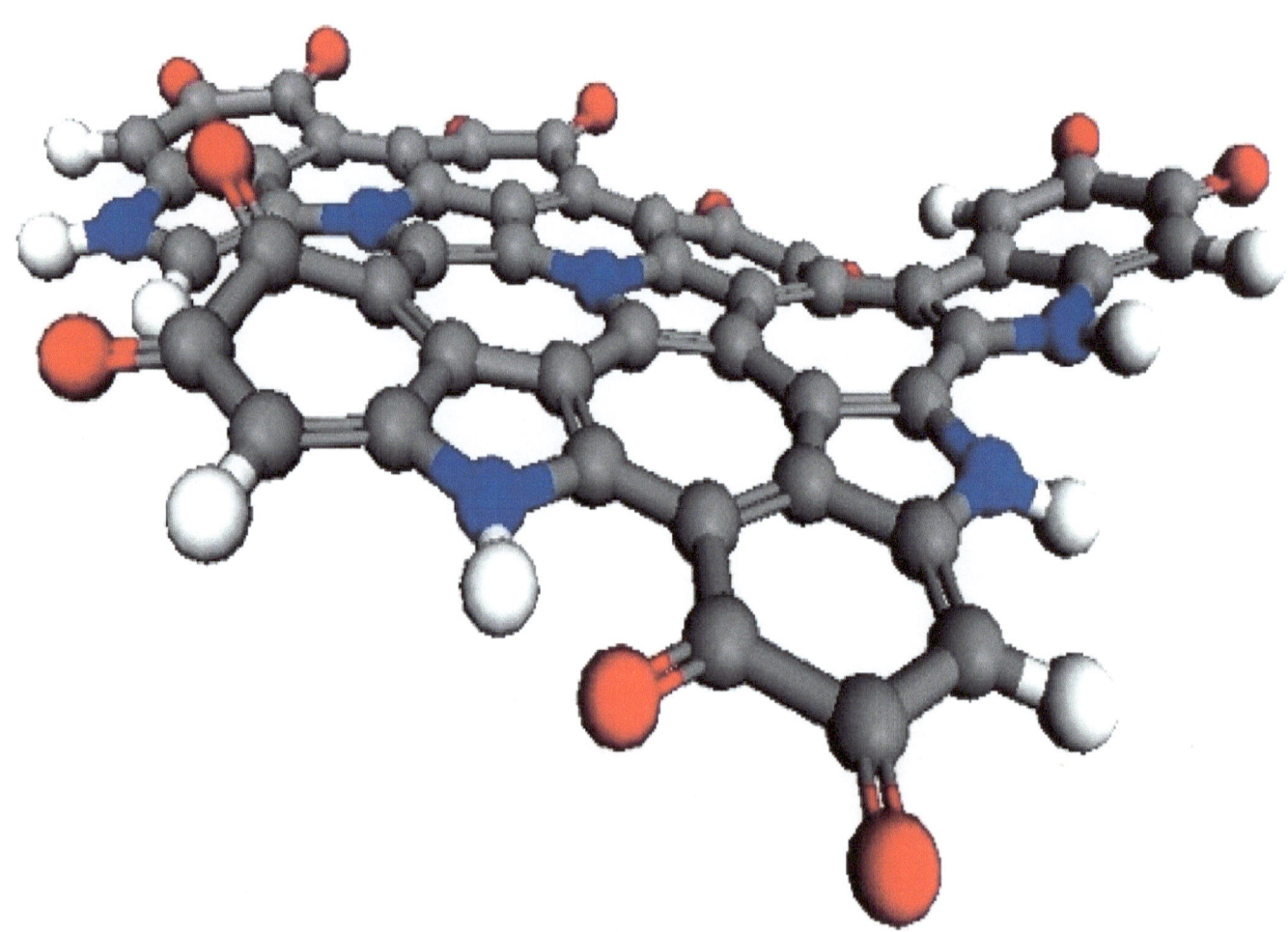

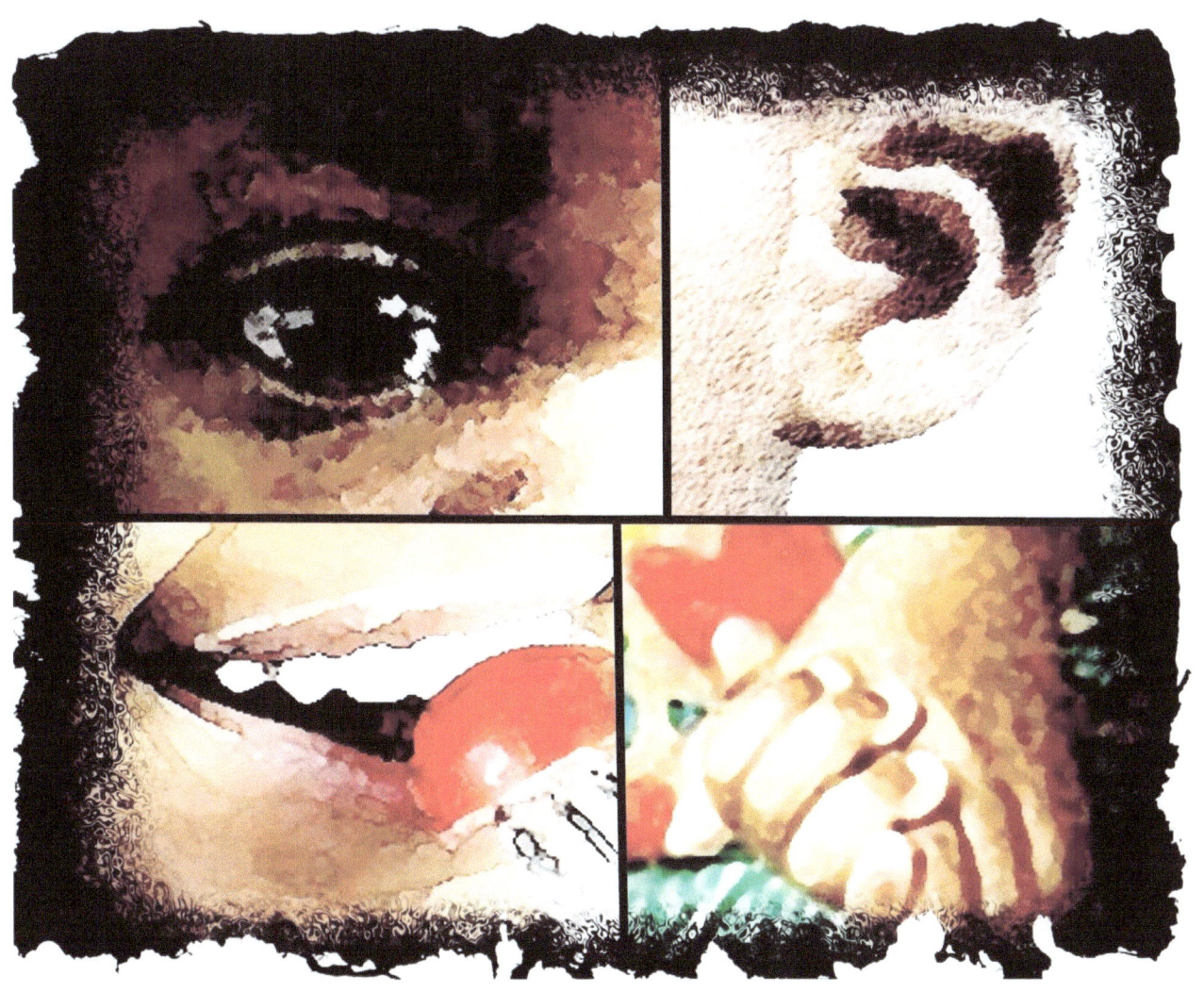

I can see what I hear
and taste what I touch

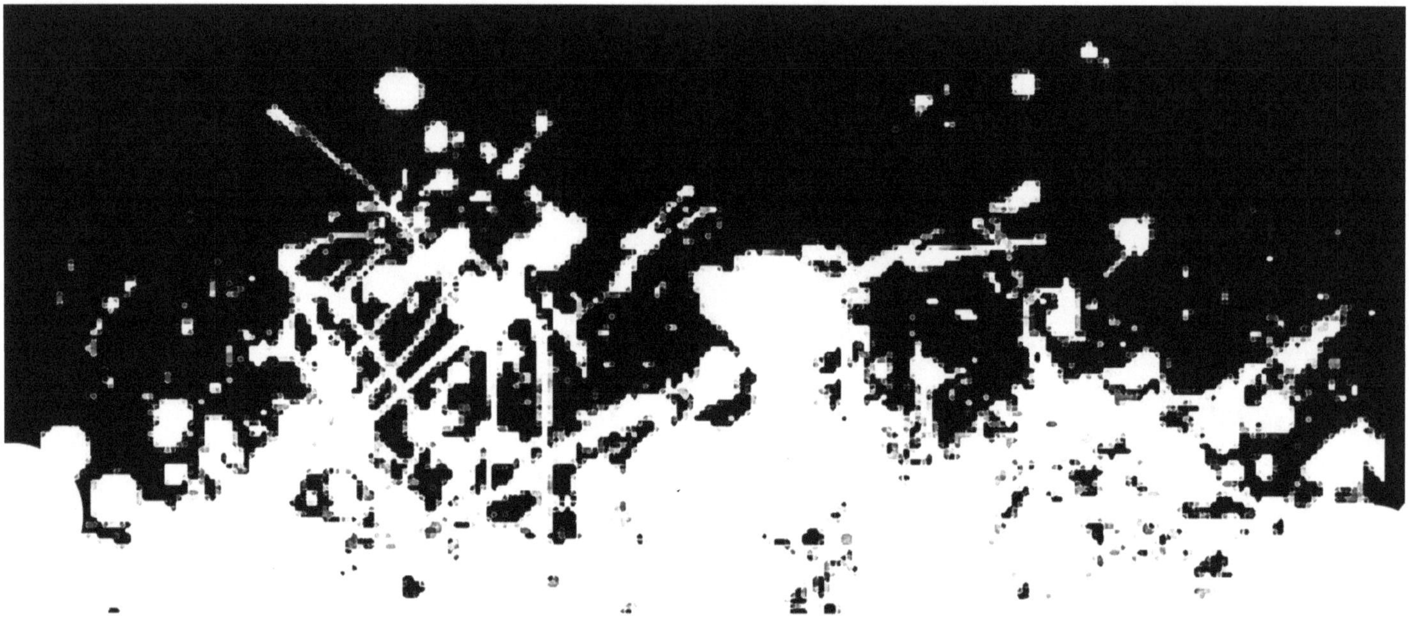

**The more Melanin you have
the more your senses can do**

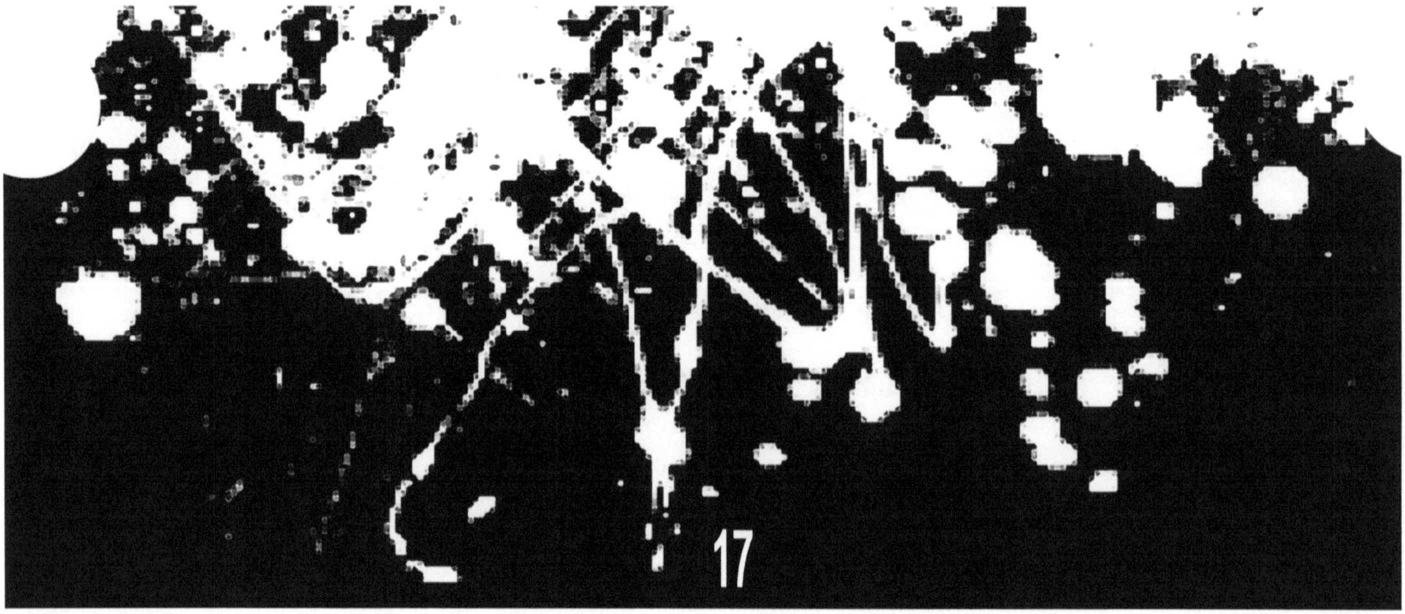

17

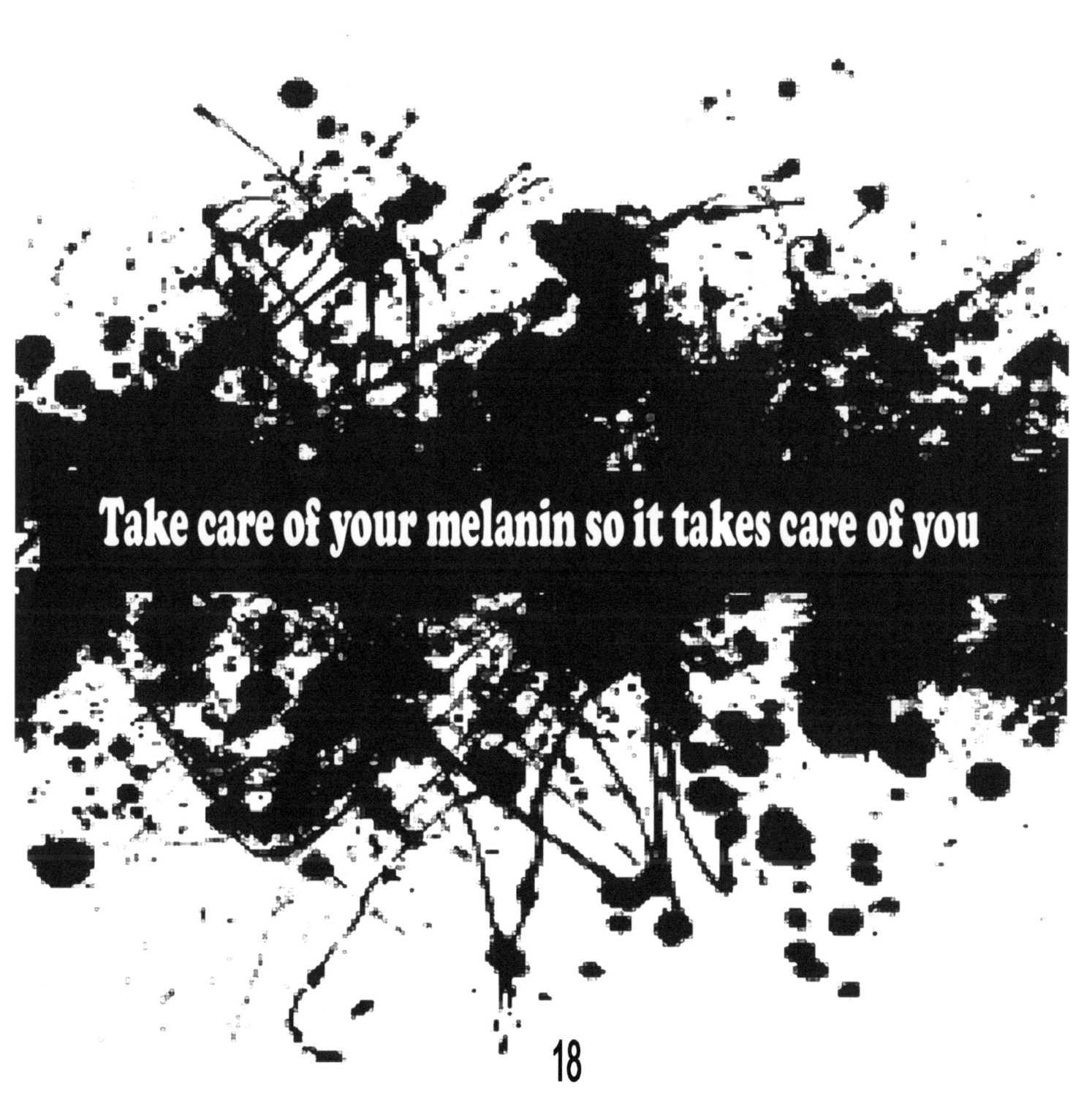

Take care of your melanin so it takes care of you

18

MELANIN
AND YOUR SENSES

WHAT IS MELANIN?

Melanin is a dark substance found throughout nature, in plants, animals and people. The word melanin comes from the Greek word melanos which means black. One of the oldest African terms for black is the Kemetic or Egyptian word kem. Ancient Africans studied blackness or melanin ages ago and have provided a lot of information that is still being studied today. After several hundred years of researching the mysteries of melanin, western scientists finally agree that melanin is the chemical key to life and living. Melanin is a vital link that connects all of creation. This carbon based substance is found in space among the planets and stars and all over the planet earth. Melanin is biological living light. It is an effective absorber of light and heat from the sun and is capable of changing energy into various other forms of energy. Melanin exists throughout the human body and acts as a conductor that helps organs and systems with their functions. It is the substance that gives the skin, hair, and iris of the eye their natural color. Melanin makes it possible for the senses of the body to do their best work. The more melanin you have the more your senses can do.

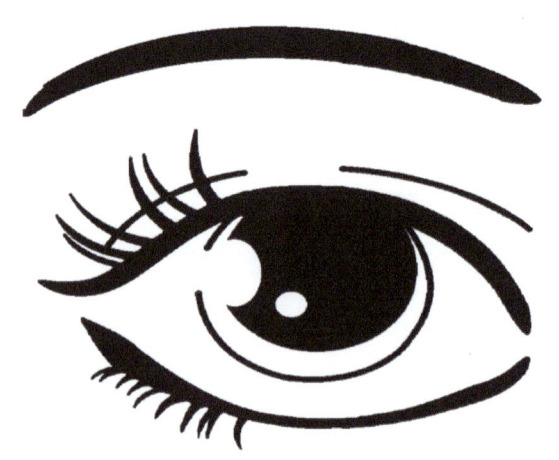

THE EYES HAVE IT

23

Melanin inside of our eyes helps us to see the full range of color spectrum. All human eyes have retinal melanin in a layer just below the surface of the rod and cones. Rods are sensory organ receptors that produce black and white vision by literally capturing light when it passes through the pupil and reaches the retina. The retina is a thin sheet of cells inside the back of the eyeball. When the nerves go from the cells of the retina to the back of the cerebrum (the largest part of the brain), this connection makes it possible for you to see.

 # DO YOU HEAR WHAT I HEAR?

25

The ear is responsible for the sense of hearing and balance. There are three major sections of the ear, the outer ear, middle ear and inner ear. The middle and inner ears are located inside of the head. Tubes go from the side of your head to your eardrums. Each eardrum is a thin piece of skin that vibrates when sound comes to it. Hearing occurs when our ears change this movement of air into sound. Melanin in the ear helps us to hear every detail of that sound.

A TALE OF
TWO SENSES

WORKING
TOGETHER

27

Our ability to smell occurs as we breathe in and air travels into the nose. Cells in the nose sense tiny things in the air. The cells send messages to the brain and the brain tells us what kind of odor it is. Our sense of smell is well connected to our sense of taste and helps to make tasting possible. Just as we need a nerve pathway to the brain to help us smell, we need the same process to help us taste. Tiny taste buds on the tongue and in the mouth work together to tell us about taste. Four major taste buds tell us if something is bitter, salty, sour, or sweet. When we eat food we taste and smell all at the same time. The presence of melanin helps these senses to work in a great way. Melanin increases sensitivity in the nose and mouth and helps to strengthen the signals to the brain.

TOUCHING...

Melanin in the skin serves and protects. It protects the body from dangerous UV rays from the sun and determines skin color. It also serves as an enhancer to the sense of touch.

In the skin and outside the body there are millions of nerve endings. There are five different kinds of nerve endings that tell the brain about five different feelings, touch, heat, cold, pressure and pain. There are extra nerve endings inside of the fingertips. These nerve endings are very sensitive and help us feel every sensation of touch.

CEREBRUM

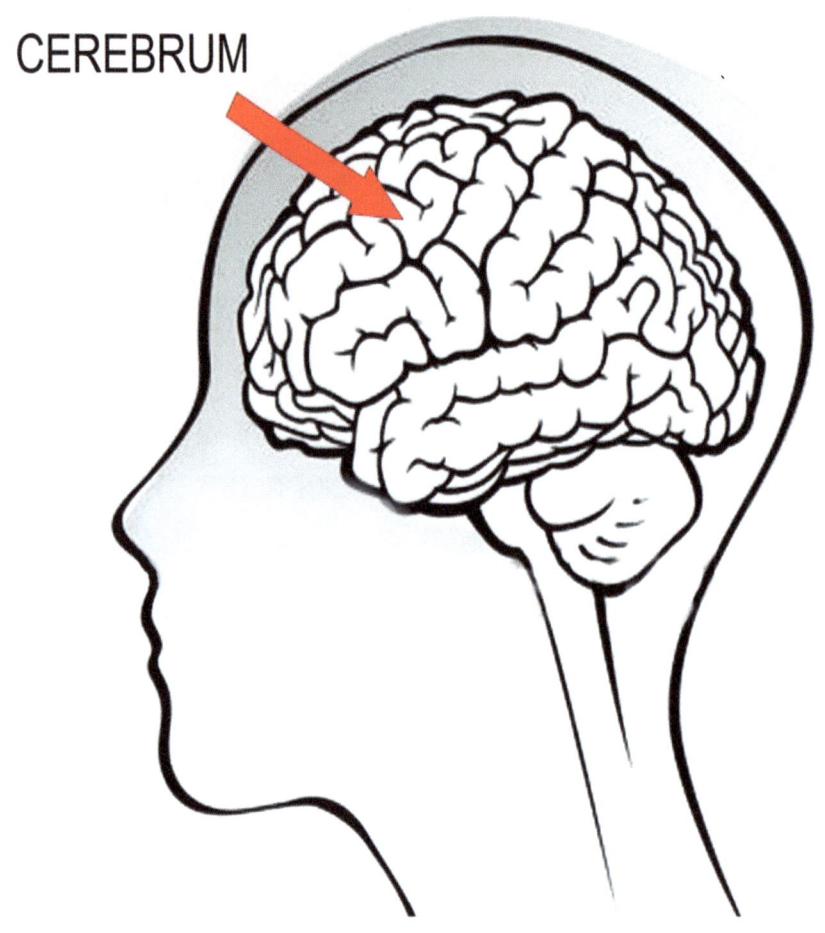

The cerebrum is the largest part of the brain. Nerves take impulses to the cerebrum and tell the cerebrum about sound, light, taste, and touch.

References

Africa, L., 2009, Melanin: What Makes Black People Black. Long Island, NY; Seaburn Publishing Group

American Journal of Physiology ; November 2006 vol. 291

King, R.D., 3001. Melanin: A Key to Freedom. Chicago; Lushema Books

Systems and Functions- Human Body, 1979. McDonald Publishing Co.